T0123640

essentials

essentials liefern aktuelles Wissen in konzentrierter Form. Die Essenz dessen, worauf es als „State-of-the-Art" in der gegenwärtigen Fachdiskussion oder in der Praxis ankommt. *essentials* informieren schnell, unkompliziert und verständlich

- als Einführung in ein aktuelles Thema aus Ihrem Fachgebiet
- als Einstieg in ein für Sie noch unbekanntes Themenfeld
- als Einblick, um zum Thema mitreden zu können

Die Bücher in elektronischer und gedruckter Form bringen das Expertenwissen von Springer-Fachautoren kompakt zur Darstellung. Sie sind besonders für die Nutzung als eBook auf Tablet-PCs, eBook-Readern und Smartphones geeignet. *essentials:* Wissensbausteine aus den Wirtschafts-, Sozial- und Geisteswissenschaften, aus Technik und Naturwissenschaften sowie aus Medizin, Psychologie und Gesundheitsberufen. Von renommierten Autoren aller Springer-Verlagsmarken.

Weitere Bände in der Reihe http://www.springer.com/series/13088

J. Matthias Starck

Scientific Peer Review

Guidelines for Informative Peer Review

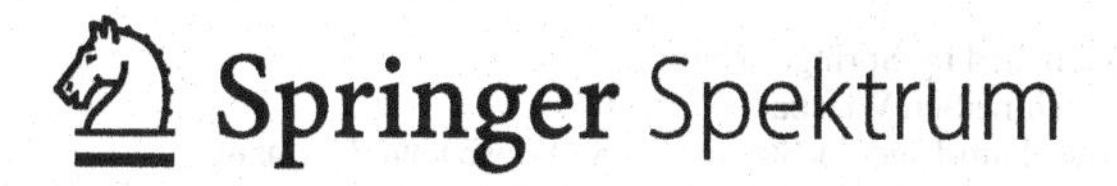

Prof. Dr. J. Matthias Starck
Department for Biology II
Ludwig-Maximilians-University
Munich (LMU)
Planegg-Martinsried, Germany

ISSN 2197-6708 ISSN 2197-6716 (electronic)
essentials
ISBN 978-3-658-19914-2 ISBN 978-3-658-19915-9 (eBook)
https://doi.org/10.1007/978-3-658-19915-9

Library of Congress Control Number: 2017955536

Springer Spektrum

Printed on acid-free paper

This Springer Spektrum imprint is published by Springer Nature
The registered company is Springer Fachmedien Wiesbaden GmbH
The registered company address is: Abraham-Lincoln-Str. 46, 65189 Wiesbaden, Germany

What to find in this *essential*

- A short analysis of basic principles how science works, how science is communicated, and how it is published.
- A comprehensive presentation of how to write a review for a scientific journal.
- A discussion of ethical guidelines applicable in scientific publishing.
- An analysis and critique of various peer review procedures so that inherent problems become transparent, alternative options are recognized, and possible solutions can be found.

Contents

1 Introduction

Peer review is the critical assessment of scientific[1] reports by independent experts. It scrutinizes manuscripts, grant proposals, or job applications for the correct application of the principles of science, correct scientific methodology, presentation according to the standards of scientific publishing, originality of research, and for legal and ethical correctness. It is the widely accepted safeguarding mechanism aiming to secure that research has been carried out appropriately, that hypotheses are phrased clearly, methods appropriate, results presented correctly, and all possible interpretations considered. Peer review gives authors feedback to improve the quality of their research papers before publication and helps editors decide which manuscripts should be published—it represents mutual collegial engagement of scientists, not judgment.

To write good, informative, and fair reviews, we need to understand the basic principles of how science works, the principles of scientific methodology, and how science is communicated.[2] By understanding the processes and potential challenges of the peer review process, we may be able not only to perform well in the existing system, but also to participate in and contribute to its further development and improvement.

[1]Science/scientific always and exclusively refers to "natural sciences" as distinct from humanities and mathematics.

[2]This essential analyses and discusses the general principles of what is communicated, not linguistic aspects of communication. This important aspect is comprehensively covered by Paltridge (2017).

J.M. Starck, *Scientific Peer Review*, essentials,
https://doi.org/10.1007/978-3-658-19915-9_1

How Science Works

2

2.1 Philosophical Principles in Science

All modern, natural sciences follow a deductive process. From observations of the world around us, hypotheses are phrased that allow for certain predictions. Hypotheses are tested using appropriate methods, and the results are interpreted on the basis of existing knowledge.

The critical rationalism[1] (e.g., Popper 1935) holds that science proceeds by falsifying existing hypotheses and that hypotheses cannot be verified. Science proceeds by rejecting hypotheses and replacing them by alternate, yet untested hypotheses (that present more plausible explanations in the light of the results of the study). If the results of a study do not reject the existing hypothesis, the hypothesis (H_0) persists, but one cannot be sure that it is correct—it just represents the currently best available explanation until it is rejected and replaced by a better one. Failure to reject a hypothesis must not be interpreted as verification of that particular hypothesis; it is possible that either the methods to test the hypothesis were inappropriate or the alternate hypotheses may have been incorrect.

Tests, i.e., experiments or comparisons, are designed in the context of existing knowledge and they provide empirical data which allow for explanations. Ideally, tests are experiments and provide proximate causal explanations. However, in biology tests may also be comparative or correlational. Indeed, among the natural sciences, biology is special because explanations in biology have an

[1]Of course, other philosophical concepts of the foundations of science exist and critical rationalism is not without critique. However, until today it still stands as the most widely accepted concept explaining how science works.

J.M. Starck, *Scientific Peer Review*, essentials,
https://doi.org/10.1007/978-3-658-19915-9_2

ultimate (historical, evolutionary) and a proximate (mechanistic) component (Bock 2017). Ultimate, historical, evolutionary causations require comparative/correlational methods and a phylogenetic tool-set. Proximate causations can be tested by experiments. Because organisms are inevitably historical (descendent by evolution) and functional, biological research is always concerned with both, ultimate and proximate explanations.

2.2 Principles of Scientific Methodology

> What is absolutely essential in the scientific methodology is not that empirical observations are made, but that these observations are objective as opposed to subjective—hence the term objective science. Objective empirical observations in the philosophy of science means that the same observations can be made by any person having the abilities to do so (Bock 2007).

An implicit corollary of the philosophical concept of science is that it requires communication among scientists. Only when hypotheses, tests, and results are communicated are they open to further testing by independent researchers and thus scientific progress is possible. This characterizes a scientific methodology which became established as standard among researchers in natural sciences. Over the centuries, this methodology has been formalized in publications (among other forms) and principles of communicating science have been established. Some of these principles have been consolidated as laws (e.g., copyright) others are rather "industrial standards" of the community, i.e., generally accepted and practiced, but open to challenge and development.

The pillars of scientific methodology are *reproducibility*,[2] *transparency*, and *honesty*! *Reproducibility* means that each study must be documented so that anybody else can precisely replicate the exact study. *Transparency* refers to the complete and correct documentation of the used materials and methods, including access to materials. *Honesty* refers to the complete and detailed documentation

[2]**Reproducibility** refers to the variation in measurements made on a subject under changing conditions, i.e., different measurement methods or instruments being used, measurements being made by different observers, or measurements being made over a period of time. In contrast, **repeatability** refers to the variation in repeated measurements made on the same subject under identical conditions. This means that measurements are made by the same instrument or method, the same observer (or rater) if human input is required, and that the measurements are made over a short period of time.

of a study so that all necessary details are given that are needed to reproduce the study. Science is based on honesty because fraud at the lab-bench level is practically impossible to detect.—Maintaining the integrity of science requires the collaborative effort of all parties involved: researchers, institutions, funding agencies, legislation, and journals—and it requires additional tools for peer review.

2.3 The Ethical and Legal Framework of Science

Science per se has no moral and no law—however, for the welfare of humans and the world around us, science has to proceed within the legal and ethical framework defined by humans. This framework is usually represented by the national law of the country in which the research is conducted and science must proceed within the (national) framework of copyright, animal protection, ethical rules for animal experimentation, protection against discrimination, and protection of ethnical minorities. The disparity of nations results in an according diversity of legal regulations.—Some journals/editors therefore also request respect for international agreements (e.g., International Whaling Commission; Convention of International Trade in Endangered Species, CITES) to avoid the relaxed conditions of some nations.

2.4 Science in the Real World

Scientists are embedded in a societal and political environment and the governments of most industrial nations have set goals for science, research, and innovation. By providing goals, guidelines, and funding, governments actively direct how science is organized and how science is communicated. For example, the European Commission calls for *open science*, which entails the ongoing transitions in the way research is performed, researchers collaborate, knowledge is shared, and science is organized (Walsh 2016). By promoting '*sharing knowledge as early as possible*,' politics actively aim at pre-review publication thus modifying communication of science. By setting political goals also on how science is organized, politics engage actively in the development of science and the societal context of science (European Commission 2008; Pan and Kalinaki 2015).

It would be naïve and not reasonable to assume that science would be free. Scientists are responsible for following the principles of science and for proceeding within scientific methodology, but they are also embedded in an ethical, legal, social, and political environment that contributes to the validation and acceptance of science.—We need to understand the processes in the system, the different rules a person can take in the system, the responsibilities, and the limitations of the system to act responsibly as members of the community and to potentially contribute to its further advancement—or, as Csiszar (2016) stated: "*[…] Peer review did not develop simply out of scientists' need to trust one another's research. It was also a response to political demands for public accountability. […]*"

3 Essence of Scientific Communication

Over the centuries, scientific communication has evolved as publishing scientific papers (among other formats and with some heterogeneity among fields of science). Scrutinizing the process of scientific publishing to its essence, it comes down to *documentation*, *validation*, *publishing*, and *archiving*.

Scientific journals publish deductive, hypothesis-based research. However, a number of journals also accept descriptive or methodological studies that are not explicitly hypothesis-based. Principally descriptive work is non-scientific, technical information, but it can be complicated and important for the community. The same holds for "negative results," i.e., studies that fail to falsify a hypothesis—they can be technically demanding and of high informative value for the scientific community. Therefore, aims and scopes of a journal characterize what type of manuscripts to publish and what the criteria for possible acceptance are. However, criteria for technical and descriptive studies are always difficult because they very much depend on the validation by the editor.

3.1 Documentation

Documentation is the presentation of scientific work as a manuscript. A scientific paper must document all ideas, hypotheses, materials and methods, results and interpretations related to the presented study. This results in a logic structure of a paper consisting of: (i) an "Introduction" that presents observations (including previous work of others) and hypotheses (see Sect. 2.1); (ii) a "Materials and Methods" section that provides all details on the used materials and methods (see Sect. 2.2), including legal and ethical clearing of a study (see Sect. 2.3), information on access to material and voucher specimens so that anybody who is willing to repeat the

J.M. Starck, *Scientific Peer Review,* essentials,
https://doi.org/10.1007/978-3-658-19915-9_3

study can do so; (iii) a "Results" section that reports results of the study, and finally a (iv) "Discussion" that presents deduced interpretations of the results and places them in the context of what is already known. The documentation of research in a scientific paper should mirror the logic of science and not the researcher's personal history with his/her research topic.

Most scientific journals request that the presented research be *original* and *novel.* From a pure scientific point of view, this is not necessary, not even expected, because science should be reproducible, i.e., reproducing published experiments (and potentially confirming previous results) is part of the scientific process.

3.2 Validation

Validation, i.e., peer review, is the critical assessment of content and style of a manuscript by independent experts (peers). It requires comprehensive knowledge about how science works, how it is documented, and it requires expert knowledge about the specific field of a study. Peer reviewers must be independent of the authors as well as of the editorial staff of a journal. Their written evaluation provides an important expert opinion that helps the editor to decide about the publication of the manuscript. The editor will ultimately decide whether the manuscript fits aims and scope of the journal, meets the scientific quality standards, and that what is reported is ethical, accurate, legally approved, and relevant to the readership.

3.3 Publishing

Publishing makes contents available to the community/public. It adds value to the content by editing, formatting, administrating, distributing, and providing access to contents and supplemental materials. The publisher provides the structures for the submission of manuscripts, their validation, and the editorial processes. Publishing is a business, thus, by adding value to content, the publisher makes an economic profit. In scientific publishing, economic interests must not affect published content, thus publishers must not influence the published content (of a scientific journal).

3.4 Archiving

Scientific publications are documents, i.e., they must be available, retrievable, and archived in an unchangeable form, theoretically for an unlimited period of time. Thus, each publication requires a unique identifier. Traditionally, these were author(s), year, journal, volume and page numbers, but recently these have been replaced by the DOI (digital object identifier). Publishers and librarians together play a pivotal role in archiving and retrieving published material—in principle for centuries.

4 Why Peer Review?

The philosophy of science does not request quality control; the ideal scientific paper is clear, straightforward, and correct. Also, scientific methodology or the legal and ethical framework for research do not request a safeguarding mechanism. If all research were reproducible, transparent, documented honestly, and conducted in respect of copyright, animal protection, ethical regulations of experimentation, anti-discrimination laws, and minority rights, no formal quality control would be necessary.—However, life is far from ideal and research is done by potentially fallible humans. Because we are not free from errors and mistakes, and because flaws may creep in research practice, peer review has been installed as a safeguarding mechanism.

4.1 What Peer Review Can Do

Peer review has been established as a tool to recognize research that excels others, and to filter out science that was ill-conceived, poorly performed, or misinterpreted. Also, the sheer number of manuscript submissions makes it necessary to recognize studies by their quality, i.e., their potential to contribute to the scientific progress. Today, peer review is the industrial standard ensuring that research has been conducted according to the principles of science, that correct methodology and documentation have been applied and that interpretations are solidly based on results. Positively speaking, the purpose of peer review is to foster collaboration on important questions and topics (Kennan 2016). However, the constantly increasing number of active researchers in the scientific community and the exponential increase in number of manuscript submissions make it also necessary to sort out studies that do not reach the necessary scientific quality standard.

J.M. Starck, *Scientific Peer Review*, essentials,
https://doi.org/10.1007/978-3-658-19915-9_4

Peer review is a collegial, thus social, process. It is founded on expert knowledge of the subject, methodology and literature relevant for specific decisions, but it is also inevitably subjective. The subjectivity of this approach could be seen as a strength as well as a weakness. If conducted fair and following recommendations of best practice, it can help assess elements of research which are challenging to quantify, e.g., novelty, it can deliver more nuanced and detailed understandings of research in the context of research production, and it can help authors improve the presentation of their science.

4.2 Limitations of Peer Review

Manuscripts are peer reviewed on the basis of the information provided by the authors. Therefore, it is practically impossible to recognize fraud and misconduct on the lab-bench level. The responsibility for research integrity is with the individual researchers.

The prevalence of fraud in science is difficult to grasp and numbers are notoriously incomplete. The few data available draw a frightening picture: in a survey among US universities, up to 50% of the scientists said they had knowledge of scientific misconduct (e.g., Swazey et al. 1993; Gross 2016). A meta-analysis based on 18 studies reported that 2.6% of scientists admitted own misconduct, but 16.6% knew of misconduct by others (Fanelli 2009). With the digitization of publishing, the number of retractions can be used as a measure of incorrectly published research, including misconduct. In absolute numbers, retractions have increased dramatically over the past 20 years, but these absolute numbers rather reflect the continuously growing field of science while the prevalence of misconduct (number of retractions per number of publications) appears to have been stable over the years (Fanelli 2013).—Nuijten et al. (2016) used a retrograde software to analyze over 250,000 statistical analyses from papers published between 1985 and 2013 and found that about 50% of the papers from 8 major psychology journals reported incorrect statistics. This study caught cases from simple typos to scientific fraud without categorization. However, the alarming message is the distressingly high number of scientific publications containing mistakes (of various kinds). Even worse, Joannides (2005) presented an analysis suggesting that most research findings are false for most research designs and for most fields because of incorrect applications of statistics.

Baker (2016) reported that "*More than 70% of researchers have tried and failed to reproduce another scientist's experiments, and more than half have failed to reproduce their own experiments [...].*" While these numbers are suitable for invoking a credibility crisis in science and scientific communication, one might ask: how would these numbers look like without peer review? It needs to be highlighted that peer reviewing cannot detect fraud and incorrect documentation, but it makes an important contribution to secure complete and correct documentation of materials and methods, thus making studies reproducible. Various actions on the lab side may and must be undertaken to improve reproducibility of research.

These numbers have been interpreted as evidence that peer review is ineffective to filter out incorrect publications (see also Sect. 9.1). However, the numbers catch two different features of scientific publishing: (1) various aspects of misconduct on the one side (authors), and (2) incompetence to recognize flaws on the other side (reviewers). Are we researchers a community of dishonest and incompetent individuals? And, what is the contribution of peer review? Competent peer review can recognize only misconceptions, flawed experiments and testing, and inconclusive discussions, but it is ineffective against scientific misconduct (fraud, falsification and fabrication of data, plagiarism; e.g., Benos et al. 2007). Peer review depends on competence, trust, and goodwill of ALL participants. It makes an assumption of honesty and can assist in establishing scientific validity, but it cannot guarantee it—it is doomed to failure if the authors are dishonest or the reviewers incompetent.

4.3 Historical Note

Peer review has been practiced since 1731 when *The Society for the Improvement of Medical Knowledge*, which later became the *Royal Society of Edinburgh*, sent out scientific communications for comments by experts in the same field. Of course, at that time there was no standard of reviewing and it took centuries of development to establish a formalized and standardized peer reviewing process. The implementation of the peer review process developed in response to the increasing specialization in science. That development was unorganized and many journals produced their own reviewing systems, often dependent on the editor in charge at the time (Rennie 2003).

Peer review, as we know it today, became established only during the years after the Second World War and was institutionalized in the mid-1980s. With globalization, the industrialization of biomedical research, digitalization and the internet, the peer reviewing process became more standardized and at the same time journals started to experiment with various systems making use of the fast and unrestrained communication options provided through the internet. Today, several versions of reviewing processes coexist, each favoring different formal aspects of communicating reviews with authors, but their intellectual core, i.e., the review, is principally the same in all systems. There is little empirical support for advantages or disadvantages of these different reviewing systems, instead opinions in favor of one or the other system are strong. Before we do not have explicit empirical support for one or the other system, any of the advertised advantages of new arrangements are unsupported assertions (Rennie 2012).

The Journal Peer Review 5

Peer review scrutinizes different aspects of a manuscript. Some are defined by (1) basic principles of science, (2) scientific methodology, and (3) the legal/ethical framework in which the scientific community is embedded; others are defined by (4) the publication process and (5) the presentation of science. Peer review cannot detect plagiarism, fraud, and misconduct.—Language, journal specific format, and style of presentation are of minor concern for a reviewer.

5.1 Science

Science is the core of a paper, therefore, a review may start by considering if the presented research is in accordance with the essential principles of science. Thus, a reviewer first checks the hypotheses, the appropriateness of tests/experiments, and if the study has resulted in the rejection of a null-hypothesis.—For scientific journals, the lack of presenting a hypothesis[1] or the failure to reject the null-hypothesis (lack of statistically significant results[2]) are inacceptable because

[1]Strictly speaking, a paper without a hypothesis is not scientific. "*Nothing is known about* ..." is not a scientific hypothesis, but curiosity aiming to filling a gap in our knowledge by descriptive work. Descriptive work is important as an observation, and it may require sophisticated methods and expert knowledge, but it is not scientific because it does not contribute to the scientific process by testing hypotheses.

[2]Some colleagues call for the publication of non-significant results. However, non-significant results cannot distinguish between the lack of an effect, inappropriate methods for testing the hypothesis, or an incorrect hypothesis. Thus, non-significant results are always ambiguous. Non-significant results are important for researchers, but they are lab-bench data; they do not advance science because they do not reject a hypothesis.

J.M. Starck, *Scientific Peer Review*, essentials,
https://doi.org/10.1007/978-3-658-19915-9_5

they violate basic principles of science. The lack of an explicit hypothesis in the introduction of a manuscript may be a matter of presentation and does not necessarily mean that the paper cannot be rescued. Guidance by the reviewer may help to improve the presentation of science.—Some journals also accept descriptive papers, technical notes, or negative results (see above).

Appropriate tests and correct experimental design are key features of scientific studies. The causal relationships, i.e., why a certain experimental design is testing the hypothesis, should be explained in detail. In correlative and comparative studies,[3] the authors need to explain why a certain association between variables is thought to provide a biological explanation.

5.2 Methodology

Reproducibility One of the key principles in scientific publishing is that any scientific study must be fully reproducible for anybody who is interested and has the abilities to do so. Therefore, all materials and methods must be disclosed and made accessible to the scientific community in all necessary detail so that the study can be reproduced and tested.—Reproducibility is a fundamental aspect of science and reviewers should always comment on this.

Transparency and accessibility Recent years have seen an extensive debate about what kind of data should be made accessible to the community (e.g., Davies et al. 2017). From a purely scientific point of view, this is the information that is necessary to reproduce a study. These can be catalogue numbers of museum specimens, information about deposition and accessibility to voucher specimens, genetic sequence data, and all kinds of information required to reproduce the study. It must be raw data and original images in case of destructive methods, ephemeral subjects of the study, or temporarily changing subjects.—Beyond the scientifically necessary, a growing consensus in the scientific community requests that ALL data that were collected in a study be made available to facilitate future studies. This includes also data that could be easily collected when reproducing

[3]Correlative and comparative studies that compare different species always include a historical/evolutionary perspective, implicitly or explicitly. Thus they must report appropriate phylogenetic information and the number of species must be adjusted to the level of comparison (e.g., Garland and Adolph 1994).

the study. This is a matter of changing perception of what is good practice in academia, and it is often associated with strong opinions about data sharing. From a purely scientific point of view, reproducing a study means collection of data in an identical (experimental) setup. Data sharing provides easy access to data and saves work, but it does not create new information and does not reproduce a study—it allows repeating an analysis or using data in new analyses. If original data collected in a study are presented in a manuscript or in supplemental material, quality control of the shared data is necessary. Sharing data potentially means sharing error.

Honesty The ultimate responsibility for good research practice lies with the individual researchers. For the reviewer, it is practically impossible to detect fraud and misconduct on the lab-bench level. Also, self-plagiarism, plagiarism or recycling of (own) material can be detected only by the specialist and rather by chance. Any suspicion of incorrect research practice or unethical behavior should be reported confidentially to the editor for further investigation.—Most major journals routinely use plagiarism detect software to screen newly submitted manuscripts, so technology can help with the laborious and somewhat stochastic work individuals could do.

5.3 Legal/Ethical Framework

All research must be conducted in the legal and ethical framework of the country in which it is conducted. In addition, international legislative or industrial ethical norms may apply according to the journal's policy.—Permit and permission numbers together with the issuing authorities must always be given in the materials and methods section. Correct documentation is a matter of honesty, but the reviewer can check for complete information.

Copyright protects intellectual property and grants the creator of an original work exclusive rights for its use and distribution. The rights are usually limited for a certain period (50–100 years after the author's death—period depending on national law). Using copyright protected material without formal permission from the copyright holder and without explicit reference is plagiarism. In the reviewing process, only the specialist may possibly detect figures or text that have been plagiarized, but in general the scientific world has grown too big so that nobody can expect that an individual reviewer can detect plagiarism. Today, automated computer programs check plagiarism against anything published in the internet and provide detailed quantitative comparisons of submitted manuscripts with any text found online. The implementation of such plagiarism check programs and their correct application is the responsibility of the editorial offices. Plagiarism check programs report overall similarity of text, the interpretation of the similarity requires thoughtful and understanding analysis because similarity, if not identity, of text may be acceptable for plain technical information (description of Materials and methods, list of references), but not for introduction, results, or discussion.

The legal and ethical framework is defined by the society we live in. It is developing and changing together with the changes in the social, cultural, and political environment of the society. The ethical framework of scientific publishing requests an increasing sensitivity to bias and explicit documentation of that sensitivity. A growing number of peer-reviewed journals have installed editorial policies requiring minority, sex- or gender-specific reporting of scientific research.[4]

5.4 Publishing

The publishing process has specific requirements for manuscripts. Many of these requirements are determined by practical reasons and not by principles of science. Some have developed into a kind of industrial standard and are largely accepted, while others are journal specific style and format requests.

Originality and novelty of a study are two of these industrial standards requested from scientific publishing. Originality and novelty of research are no requests of science. In contrast, science demands that studies are reproducible and should be reproduced by others. The demand for originality and novelty stems from the limitation of space, time, and energy in scientific publishing and from the fact that only few researchers would wish to read content that is already known. It requires the expert reviewer to evaluate the degree of novelty and originality of the research. However, both are soft criteria because the degree of novelty and originality are only vaguely outlined and may be influenced by the reviewer's or the editor's knowledge. Reproducing studies should not be mistaken with plagiarism, which certainly is the down-side of reproducibility.—Most journals also request that a paper is concise. Of course, this is also a pragmatic request owing to the limited space for publishing and the readability of a study.

[4]For biomedical research the European Association of Science Editors (EASE) has developed the sex and gender equity in research (SAGER) guidelines (Heidari et al. 2016; De Castro et al. 2016).

5.5 Presentation of Science

Science is technical communication. It is simple, concise, and straightforward, but neither pathetic, complicated nor validating. In a research paper, the presentation follows a standard scheme (title, abstract, introduction, materials and methods, results, discussion, references, tables and figures, supplementary online material). The sequence of the sections may differ among journals, but it always mirrors the main principles of how science is presented.

The *Introduction* presents the (observational) background and important previous research, the hypotheses and predictions. There is no need to say how important the study is, the community will eventually decide about this—and there is no need to say that this is the first study on that subject because for most journals originality and novelty are basic requests.

The *Materials and Methods* section documents all materials and methods used to a degree that the study is fully reproducible by anybody who wishes to do so. The need for reproducibility has important implications for each scientific paper because it requests access to original material for all who wish to reproduce a study. These may be measured raw data/images if the methods of the study were destructive or the material is deciduous (e.g., Davies et al. 2017).

The *Results* section should present all results and only results from the study. Of course, presentation of results must be clear, complete, and explicit in text, figures or tables. Duplication of results in text, tables or figures shall be avoided and the representation should be as concise as possible. No references to earlier studies or published data from the literature should be made in the results section. It must be explicit and clear what is own and what was published earlier.—Some fields in biology (e.g., taxonomy, paleontology) accept a mixed presentation of own results with data from referenced studies. While this appears to be a traditional form of presentation, the praxis is confusing and not helpful to understand what has really been done and what is the work of others. A strict separation of new data and referenced data into results and discussion respectively, is a simple and straightforward way to avoid any misunderstanding about what has been done.

A deplorable habit is reference to "unpublished data" or "personal communication" either in results or discussion. Despite some journals' accepting reference to "personal communication" after written consent of the referenced colleague, they are invalid in a scientific sense because the origin of these data is not transparent, they have not been validated, and the original material is not accessible for future testing.

Modern online publishing techniques allow for presentation of extensive supplementary material, thus allowing for a complete documentation even of large studies. Supplementary online material is an integral part of a paper and can be referenced. It should be reviewed and scrutinized for correctness and scientifically sound presentation.

The *Discussion* should present possible interpretations of the results and relate them to previously published papers. The discussion should give straightforward answers to the hypotheses and questions raised in the introduction. Repeating the results is not necessary and not wanted. Discussions are often too long, winding, and overstating the importance of the results. While this may be understandable from the author's point of view, it is always the scientific community that will ultimately decide about the importance of the study. Science and its presentation is a simple process: we ask questions and provide answers, which are discussed in the context of what colleagues have studied before. It is the domain of the expert reviewers to carefully evaluate if the discussion presents the results of the study in relationship to previously published material.

Language, style, and format presentation It is the author's responsibility to present a paper in clear and concise language and in the appropriate format requested by a journal. Language may be a problem for non-native English speaking colleagues. While a statement about language presentation and format is usually requested, it is not the reviewer's duty to provide extensive language editing. Today, many professional language editing companies offer their services at reasonable prices, so that those who are not confident in their language skills can request professional help (best before submission).

5.6 Authorship

Authorship assignment is a matter of honesty, and usually it is impossible for the reviewer to recognize potential conflict. However, it is an important and potentially complicated issue that has traditionally been the source of conflicts. Criteria that define authorship were developed by Wagner and Kleinert (2011), further elaborated by the International Committee of Medical Journal Editors and adopted by numerous scientific journals. Authorship should be based on the following four criteria: (1) Substantial contributions to the conception or design of the work; or the acquisition, analysis, or interpretation of data for the work; (2) drafting the work or revising it critically for important intellectual content; (3) final approval of the version to be published; and (4) agreement to be accountable for all aspects of the work in ensuring that questions related to the accuracy or integrity of any part of the work are appropriately investigated and resolved. Authors should meet all four criteria. Those who do not meet all four criteria should be acknowledged. Acquisition of funding alone, collection of data alone, or general supervision of the research group alone do not constitute authorship. Also, each author should have participated sufficiently in the work to take public responsibility for appropriate portions of the content.

Most papers in science are multi-authored papers. Thus contributor rules need to be reported and described in an appropriate and preferably standardized manner. The "contributor role taxonomy" developed by Casrai.org is a helpful tool for standardizing descriptive terms and thus helping to make author contributions more comparable.

5.7 Recommendations

The expert opinion is important for the editor to make an informed decision about a manuscript. Together with the detailed report, the reviewer should always provide a recommendation and explicit reasons for it. It is the editor who ultimately makes a decision under consideration of ALL information available (including his/her own evaluation) pertinent for the journal. This decision may differ from the reviewer's recommendations (Sense about Science 2012).—Standard reviewing knows standard recommendations from accept to reject. However, there is a logic behind each of these recommendations, which should be considered before use.

Accept

When all criteria for a scientific publication are satisfied, when the presentation of science in the manuscript is concise and transparent, and all criteria pertinent to the journal are satisfied, the decision will be accept. It is very rare that the first round of reviews results in an "accept." In any case, the reviewers should provide positive statements (!) to all questions/criteria relevant for publication. An excellent manuscript deserves an excellent and detailed review. A positive recommendation also needs support, explanation, and evidence.

Reviewers should always be critical. It is easier to handle excessive critique than false positive reviews. Excessive critique by the reviewers will be rebutted by the authors, but false positive reviews will not. If not detected, false positive reviews may potentially result in the publication of flawed materials. If detected, everybody involved in the reviewing process will be disappointed: (1) the authors will be disappointed because their paper is rejected despite positive reviews; (2) the reviewers will be disappointed because their review is not appreciated/taken into consideration and will consider reviewing a waste of time; and (3) the editor will be disappointed because the reviews were useless, much time and energy were wasted, and everybody is disappointed. False positive reviews seriously corrupt the reviewing process.

Minor revision

This recommendation is generally made when (only) format or (minor) language changes are necessary. For the recommendation "minor revision" the science and scientific methodology must be correct. Minor revision means "no content change" necessary, i.e., no changes are necessary with respect to science, material, or documentation. Some paragraphs may require rewriting (shortening, more concise presentation, possibly literature added). Figures and graphs may require format changes, but also no change of content.—When minor revision is the decision, a paper will usually not be re-reviewed because no content will be changed.—Minor revision is a category, not a quantitative measure, i.e., there may be many changes necessary in terms of language and format and they may accumulate to a lot of work, but it still is "minor revision" because no content is changed.

Major revision

This recommendation is made when changes of content are necessary or the science is not presented correctly. There may be many reasons, e.g., sample size is too small and there is not enough presented material (add more specimens); statistics are not appropriate and require improvements; any other research needs to be added to support the results of the paper; additional figures may be necessary to support results, and conclusion or figures need to be improved or changed. The reviewer needs to be explicit about the reasons why "major revision" is recommended and how the revision can be achieved. As a major revision usually means changes of the content of a paper, re-review is necessary. As a reviewer, be prepared and willing to re-review the manuscript. Editors tend to ask the same reviewers because they know the manuscript and the critique, and because inviting additional/new reviewers possibly results in additional comments and requests for changes and may develop into an endless story. "Major revision" is a category and not a quantitative measure (see above). A paper nicely prepared and perfectly presented, but the sample size may be too small, then major revision is the correct recommendation.

Reject with the option of resubmission
Some journals offer this option, which is a bit of "very major revision." This recommendation is made when the topic is interesting, but the science in the paper is insufficient. It is recommended when the qualified assumption is made that the paper could be rescued by additional research (material added; sample size increased). The reviewer's recommendations should contain explicit statements how (and why—of course) to improve and change the paper.—A resubmitted manuscript is considered a new submission.

Reject
Rejection should be recommended when a paper is flawed, uses inappropriate methods, is not reproducible, does not provide proof of permits, or lacks appropriate ethical clearing, is incomprehensible or for any other reason fails to meet the acceptance criteria of the journal.

Reject before review
If the topic is inappropriate for the journal, a manuscript should be rejected **before** review. It is the (handling) editor's responsibility to sort out those manuscript submissions that are inappropriate for the journal. It saves time and efforts and helps the authors find the best possible outlet for their research. Of course, there is a continuous transition from appropriate to inappropriate and sometimes the reviewer's opinion might be supportive in one or the other direction. In principle, however, the decision if a paper is in line with the aims and scope of a journal should be made before review.

5.8 Checklist for a Complete Review

Recommendations for a complete review are difficult to present because of the heterogeneity of scientific articles depending on the field in science, the field-specific traditions of communication, and the different requirements of the journals. The "checklist" is intended as a mind map of important aspects to be considered in an informative and complete review; it may also help to standardize reviews so that evaluations of the work by different authors is equal, fair, and transparent.

General

- Key information about the research presented in an article?
- Is the research presented novel and original?
- Is the presentation concise?
- Is the documented material (images, tables) of appropriate quality?
- Is gender information (if applicable) consistently presented in the article (Heidari et al. 2016)?
- Recommendation to the editor?

Title and Abstract

- Is the title concise and informative?
- If only one sex is included in the study or if the results of the study are to be applied to only one sex or gender, the title and the abstract should specify the sex of the animals or any cells, tissues and other material derived from these and the sex and gender of human participants.

Keywords

- Search engines automatically search the title words of a paper. Keywords are additional search items that should NOT be redundant with title words and help find a paper using online search engines.—Search engines dislike too much keyword repetition, known as keyword stuffing, and may 'un-index' an article, making it hard to find it online.

Introduction

- Does the introduction contain explicit hypotheses?[5]
- Are hypotheses relevant and meaningful?
- Does the introduction present a clear and explicit description of the background for research?

[5] „*Nothing is known about …*" is not a valid reason to conduct a study and certainly no hypothesis to test. Curiosity certainly drives science, but it does not replace hypotheses that can be tested using appropriate methods.

- Does the introduction refer to the **important** previous studies in the field (excessive references should be avoided)?
- Is the introduction presented in a concise style? Long excursions are usually not necessary when introducing the study.
- Authors should report, where relevant, whether sex and/or gender differences may be expected.

Materials and methods

- Any scientific study must be fully reproducible for anybody who is interested and has the abilities to do so. Therefore, all materials and methods must be disclosed and made accessible to the scientific community so that the study can be redone and tested.
- Are the applied methods (including statistics) appropriate to test the hypotheses?[6]
- Are all methods described in all necessary details?
- Has the origin of the material as well as the depository (where the material is deposited[7]) been given in full detail (e.g., catalog numbers for museum specimens; digital data depositories[8])?
- Are species and taxonomic author given correctly?

[6]Some researchers do not feel comfortable with statistics. If you do not feel competent to consider the statistics or parts thereof, please report to the editor (you can address this in a confidential note to the editor).

[7]All material of a study should be made available to the community, ideally by deposition in a university collection or public museum. Private collections are highly problematic because they are not accessible to the public and access to material can be regulated by the private owner. Also valuable material may potentially be sold to undisclosed business partners, thus be withdrawn from public access.

[8]Any study should be reproducible, i.e., all information must be given to reproduce the study. Each study requires individual evaluation. E.g., image stacks from µCT-imaging do not necessarily need to be deposited in public depositories if species and specimens are available to reproduce the study. However, if the studied specimens were unique or are not otherwise available, or if data cannot be created again (e.g., ecological data in an ecological study), then we request raw data to be deposited either in a public depository or as supplementary material for the paper.

- Are necessary permits and ethical clearing given (collection permit, animal experimentation)?
- Is the sample size (for each method applied) made explicit?
- Is the sample size sufficient to conduct tests and support conclusions?[9]
- Have pseudoreplications been avoided?
- Are sex, gender and minority issues considered appropriately?

Results

- Do results and discussion relate to the hypotheses presented in the introduction?
- Does the results section report data from all used methods?
- Are all used methods described in the Materials and Methods section?
- Are all results appropriately documented by text, figures, graphs, and tables?
- Is the image quality appropriate?
- Do images show the details reported in the text?
- Are image labels complete, all labels explained?
- Are results presented in a non-redundant manner?[10]
- Are results presented in the most concise manner?
- Can details for a specialist be moved to an online appendix?[11]
- Where appropriate, data should be routinely presented disaggregated by sex and gender. Sex- and gender-based analyses should be reported regardless of positive or negative outcome. In clinical trials, data on withdrawals and dropouts should also be reported disaggregated by sex.

[9]Of course, sample size can vary, but even in qualitative morphological comparisons among species a sample size of one individual per species cannot (!) be accepted because it ignores interindividual variation. Only in few cases where extremely rare material is studied might a sample size of one be considered.

[10]I.e., data should NOT be repeated in text and tables; once is enough.

[11]An online appendix is an integral part of the paper and therefore can be cited. All material presented in an online appendix is equally protected by the author's copyright.

Discussion

- Does the discussion refer back to the hypotheses and does it provide explicit explanations?
- Are all possible explanations discussed, without bias for the author's preferred solution?
- The discussion should place the results in the context of previous results. Are all important publications considered?[12]
- The potential implications of sex and gender on the study results and analyses should be discussed. If a sex and gender analysis has not been conducted, the rationale should be given. Authors should further discuss the implications of the lack of such analysis on the interpretation of the results.

Supplementary Material

- Have you checked and reviewed supplementary online material?

5.9 Formal Structure of a Peer Review

Journal peer review has no formal structure and reviews are returned in all kind of formats and logical arrangements. The typical discourse structure of a journal peer review, however, has four "moves," i.e., (1) summarizing statement regarding suitability for publication, (2) outlining the article, (3) points of criticism, and (4) conclusions and recommendations (Paltridge 2017). Most journals separate information communicated to the authors and to the editor. In particular, the recommendations, though important part of a review, should be addressed to the editor and not to the authors. Many journals also provide checklists with specific questions to the reviewers that allow for a quick overview.—Ultimately, the information content of a review is the important part, but a formal structure (e.g., see checklist above) may help provide consistent, fair, and transparent reviews.

[12]Given the fact that reviewers are experts in the field of the manuscript, it may be a legitimate request from the reviewer to include their own publications with important contributions to the field. However, this is a sensitive topic because it may also be abused by the reviewers to boost up citations of their own work.

Language is a complicated issue, and the way how information (e.g., critique) is communicated may differ tremendously among people, depending on cultural background, training, language skills. Writing reviews requires different language skills than writing scientific papers and non-native speakers may generally be at disadvantage as compared to native speakers. Expressing critique in a polite manner can be a challenging task, and it has been found that non-native speakers may be blunter with their critique than native speakers. Involvement, general approval, and indirectness of speech are most frequently used as politeness strategies in journal peer review, linguistic skills that may be more difficult to use for the non-native reviewer. For all reviewers, it may be helpful to even think about linguistic strategies of expressing critique in a polite manner before writing the review (see Paltridge 2017 for a detailed and interesting analysis of the subject).

5.10 Editorial Decision

Decisions about a manuscript are made by the editor-in-chief, not the reviewers. The reviewers prepare expert recommendations as the basis for editorial decisions. However, journals should explicitly outline the criteria for decision to make the decision process fully transparent. A published list of criteria will not only help the authors to understand the editor's decision, but also the reviewers to prepare reviews so that the editor receives exactly the information he needs. Although these editorial criteria should always reflect the basic principles of how science works and how science is communicated, they may differ among journals depending on the journal's policy. Decisions may be based on expected scientific impact, degree of advance, novelty, or being of specific/general interest. Such criteria are soft and difficult to characterize because they depend on the personal knowledge and experience of the responsible editor. However, many journals are (still) limited by print space and they need to establish priority rules for the acceptance of manuscripts. For many (if not most) journals soft criteria are important to decide about the published content.[13]

[13]Few journals like PeerJ or PlosOne explicitly exclude soft decision criteria and accept all manuscripts that report technically accurate studies.

6 Responsibilities of the Reviewer

Being invited to review a manuscript is a privilege because the reviewer gets to see research results ahead of others, and because the reviewer is recognized as an expert in a field. However, accepting to review a manuscript for a journal is also a commitment that incurs a number of responsibilities towards authors and editor.

6.1 Reviewer Responsibilities Toward Authors

Peer review is the expert opinion in a scientific field—therefore, scientific expertise is a prime responsibility of reviewers towards authors. If a reviewer does not feel competent to review a paper or parts of it, the editor should be informed immediately so that the right experts can be solicited. A self-critical attitude about the expected expertise is part of the collegial honesty that is required at so many moments during scientific publishing.

Most journals practice a confidential reviewing system. Maintaining the confidentiality of the review process, i.e., not sharing information, discussing with third parties, or disclosing information from the reviewed paper, is the prime responsibility of the reviewer. If the journal practices open review and publishes the reviews together with the paper, it will inform the reviewer upon reviewing request.

In times of increasing pressure to publish, the timeliness of a review is an important responsibility of the reviewer. Reviews, providing written, unbiased feedback on the scholarly merits and the scientific value of the work in a timely manner, together with the documented basis for the reviewer's opinion, should be returned within the agreed period of time.—If the review is delayed for whatever reason, the editor should be informed and the expected date, when the review will be finished, shall be given.

J.M. Starck, *Scientific Peer Review,* essentials,
https://doi.org/10.1007/978-3-658-19915-9_6

The reviewer shall prepare detailed, fair, and knowledgeable reviews on the ground of principles of science, documentation, and publishing. Reviews shall be written polite, but explicit in their critique; they should avoid personal or derogatory comments or unsupported criticism.

6.2 Reviewer Responsibilities Toward Editors

The reviewer should notify the editor or the editorial office immediately if the review cannot be submitted in a timely manner; usually an extension can be negotiated. It is important to keep communication flowing between the editorial office and the reviewer.

Providing a thoughtful, fair, constructive, and informative critique of the submitted work, which may include review of the supplementary material provided to the journal by the author, is the prime responsibility towards editor and author(s), of course.

The reviewer should alert the editor about any potential personal or financial conflict of interest[1] and decline to review when a possibility of a conflict exists.

The reviewer should accept only manuscripts within his/her area of expertise.

The reviewer should comply with the editor's written instructions on the journal's expectations for the scope, content, and quality of the review.

The reviewer should determine scientific merit, originality, and scope of the work; indicating ways to improve it; and recommending acceptance or rejection using whatever rating scale the editor deems most useful.

The reviewer should report any ethical concerns such as any violation of accepted norms of ethical treatment of animals or human subjects or substantial similarity between the reviewed manuscript and any published paper or any manuscript concurrently submitted to another journal which may be known to the reviewer.

Reviewers should refrain from direct contact with the author(s). Reviewing is a documented process with the best possible standards of transparency. Direct communication between reviewers and authors bypasses the editor and therefore potentially prevents an informed editorial decision. Direct communication between reviewer and author corrupts the transparency of the reviewing process.

[1]This includes positive bias! Do not provide friendly biased reviews, it is considered scientific misconduct (and suitable of corrupting the system)!

6.3 How, When, and Why to Say no to a Review Request

The answer to this question is implicitly given in the reviewer responsibilities (above). Failure to comply with any of the above mentioned responsibilities of the reviewer may be a reason for declining to review. Definitely, one should decline to review: (1) if one is not an expert in the field and cannot provide a scientifically sound review, (2) if one cannot provide a thoughtful and detailed review within an agreeable period of time, and (3) if one feels biased (positively or negatively).

6.4 How Much Time Spent with Review?

There is no standard answer to this question. It depends, of course, on the reviewer's experience, the complexity and length of the manuscript, and the number of comments and necessary corrections. A young post-doc may spend several days reviewing a manuscript, while an experienced senior researcher may spend just a couple of hours for the same manuscript.—In any case, there should be enough time to provide a detailed and thoughtful review.

6.5 How Much Detail?

Again, there is no standard answer to this question. In principle, a review should cover ALL aspects of a paper (see Sect. 5.7) and every reviewer should explain and support his/her judgments. Thus, a good review can be considerable work. Some reports are submitted with only short comments such as "*This is an inferior manuscript. Reject*" or "*Great work. Publish*" with no further explanation—such reviews are insufficient and most certainly will not be considered by the editor. Such "one-liners" irresponsibly delay the reviewing process because additional reviews need to be invited thus postponing the decision.—A review should always provide the supported expert opinion about the science presented in a manuscript with appropriate recommendations for the editor. However, a reviewer does not need to correct details of language or even provide copyediting. Language editing may be provided as a courtesy towards the authors, but it is neither necessary nor expected. It is the authors' responsibility to submit the manuscript in clear, concise, and comprehensible language, and it is the publishers' responsibility to provide copyediting.

7 Ethical Rules in the Reviewing Process

Writing scientific papers, reviewing, and publishing are based on honesty. Misconduct on the lab-bench level is almost impossible to detect for reviewers and editors. Screening of manuscripts using plagiarism check software and statistical monitoring programs provide important tools to the editorial staff to detect misconduct and may help highlight cases in which more detailed investigation may be necessary. However, the confidential and sometimes non-transparent reviewing system makes the system also vulnerable to misconduct by reviewers and editors.

7.1 Reviewer Misconduct

The industrialization of science and the constantly increasing pressure on researchers to be productive (counted in the number of published papers and collected impact points) increases the overall workload and the competition among researchers. Because of increased workload, reviews may be too short, uninformative, and delayed. While one would not consider this reviewer misconduct, the transition between a poor or uninformative review and reviewer misconduct is fluent. When the peer reviewer is a potential competitor of the authors, conflict may arise when reviews are *purposely delayed* to change priorities or *unduly negative* to prevent publications by competing research groups. *Breaking confidentiality*, *plagiarizing data* from reviewed manuscripts, or *stealing ideas* are also possible forms of reviewer misconduct. If the authors recognize any form of reviewer misconduct, they should express their claims to the editor who is responsible of investigating and solving problems (see below).—Of course, *false positive reviews*, i.e., supporting friends and close colleagues by providing (unsupported) positive reviews, are equally considered reviewer misconduct,

J.M. Starck, *Scientific Peer Review*, essentials,
https://doi.org/10.1007/978-3-658-19915-9_7

although neither authors nor reviewers will complain about this as a conflict. False positive reviews are a nightmare because they are difficult to recognize, require intensive investigations by the editor, and are suitable to corrupt the entire scientific publishing system.

Editors are supposed to be independent, fair, and supportive of the scientific community and specifically their authors. Unfortunately, cases have been reported where editors were involved in misconduct. Unfair, biased, and non-transparent handling of manuscripts, breaking confidentiality, and plagiarizing data have been reported, though these cases appear to be exceptional. If the editor is involved in misconduct, only external committees and neutral institutions can investigate and help (see below).

7.2 Conflict and Solutions

Conflict between authors and reviewers, and potentially between authors and editors, requires careful handling, thorough investigation, and independent solutions. Generally, conflict between authors and reviewers will be handled by the editor, best in consultation with independent external committees maintaining transparency and confidentiality of the investigation. While reviewers are expected to write fair, informative, and collegial reviews, authors are encouraged to report any evidence of reviewer misconduct to the editor.

Independent institutions like the Committee on Publication Ethics (COPE, see Sect. 11.1) provide support and solutions in all conflict cases; COPE also elaborates workflow suggestions for formalized handling of various conflict cases. COPE publishes (anonymized) conflict cases and their solutions so that all parties (authors, reviewer, and editor) have the option of learning and understanding solutions to conflicts. There are other independent committees and national institutions that recommend best practice in scientific publishing and offer support and mediation in conflict cases (see Chap. 11). The important point is that nobody is alone and there is independent and neutral help and support when conflict arises.

The Dark Side of Scientific Publishing 8

The enormous pressure on scientists to publish on the one side and the constantly increasing workload on the other side (including peer review) bear the potential to deteriorate or even collapse the peer reviewing system. The increasingly harsh and competitive research environment has opened space for abuse of the scientific publishing process.

8.1 The Tragedy of the Reviewer Commons/ Cascading Peer Review

The exponentially increasing number of manuscript submissions to all journals, the increasing pressure on authors to publish in high ranking journals, and the overvalued importance of the impact factor of some top journals result in high rejection rates for most internationally recognized scientific journals. Rejection rates above 60% are standard and top journals have rejection rates around 90%.

Understandably, authors tend to submit their manuscripts to the highest ranking journal. Unfortunately, some authors view anonymous peer review as a stochastic process and, if the manuscript was rejected by one journal, try it with the next, hoping a new reviewer may give a different evaluation. Once a manuscript is rejected, it will cascade down with submissions to journals with a lower impact factor. This results in multiple submissions of the same manuscript to several journals until it is ultimately accepted by a journal of appropriate rank. Along with submissions, the manuscript is reviewed multiple times, resulting in an estimated average of 5–10 reviews per manuscript (Hochberg et al. 2009). Cascading of manuscripts adds to a serious overload of researchers with review requests, unnecessary effort on repeatedly rejected manuscripts, and potentially declining

J.M. Starck, *Scientific Peer Review*, essentials,
https://doi.org/10.1007/978-3-658-19915-9_8

quality of reviews. Indeed, it would be the author's responsibility to submit the manuscript to a journal of appropriate impact, but recommendations to overcome this "*tragedy of the reviewer commons*" (Hochberg et al. 2009) by mutualism, altruism, and good scientific practice are probably futile in an industrialized scientific world.—To overcome at least some of these problems, publishers have invented the "portable review," i.e., manuscripts with proven scientific merits but not being accepted in their top journals are transferred together with the received reviews to other journals (of the same publisher) that might be interested in the paper. Some of the large publishing companies, journals, and individual researchers have also agreed on the San Francisco Declaration of Research Assessment (DORA) with the aim to "*greatly reduce emphasis on the journal Impact Factor as a promotional tool*" and to provide a more content oriented assessment of scientific work. In the long term this may reduce cascading peer review.

8.2 Fake Journal Peer Review

Many journals offer, through their online submission systems, the option for the authors to suggest expert reviewers and exclude others where they assume conflict of interest. This practice is controversial. Proponents quote that ethically correct suggestions will nominate peers who are experts in the field, not involved in the study, not collaborating with the authors, at a different institution, and in no way biased. If handled correctly, these suggestions may be helpful for everybody because they enable the editor to find expert reviewers quickly who supposedly return informative reviews. In contrast, opponents point out that peer reviewers should always be invited independently by the editor and that "recommendations by the authors" would open doors for abuse. Indeed, authors may play unfair suggesting friends and collaborators in the expectation of a reviewing advantage. Even worse, recent years have seen an increasing number of cases where authors abused the system and created fake reviewer email addresses, sometimes based on names of real researchers, so that review requests were directly diverted back to the authors who provided favorable reviews on their own manuscripts.

Such abuse of the peer reviewing system is based on several weaknesses inherent to the workflow, and the lack of precaution and control by editors who accept reviewer suggestions from the authors. There are simple work-arounds to this, but they rely on the responsibility of the editors.

8.3 Failure of the System

Journal peer review is not perfect, it relies on the honesty to act correctly with respect to science and publication ethics. It is under potential "attack" from three sides: authors who wish to get their manuscripts published at any price, reviewers providing insufficient reviews, and journals that prey for publishing fees but not for good science.—In a "field experiment of scientific publishing," Bohannon (2013)[1] wrote a fatally flawed paper on a non-existing drug and submitted it to 304 open access journals. Of the 255 versions that went through the entire editing process, the paper was accepted by 157 journals and rejected by 98. Of the 106 journals that did conduct peer review, 70% accepted the paper. With no doubt, this "experiment" is a disaster for peer reviewing and for scientific publishing. The experiment has challenged the weakness of the system, i.e., honesty, and the system has failed. As mentioned several times in this text, peer review is built on trust and it will be severely corrupted if even a single player behaves unethical.

8.4 Predators

Over the past years, a rapidly growing number of predatory journals has emerged that publish almost everything as long as authors pay a publication fee. They are predatory in a sense that they hunt for the publication fee and provide only minimal publishing service and no quality control. The Beall's list of predatory open access journals (http://beallslist.weebly.com/) provides a comprehensive overview of journals that supposedly do not follow scientific and ethical standards of publishing. Criteria for recognizing predatory journals are explicitly given in this list, so it is helpful to find a path through this constantly growing jungle.—Inversely, the directory of open access journals (https://doaj.org/) is a list of open access journals that warrant quality, peer review, and open access to publications.

[1]There are more and similar „experiments" that document potential failure of the system. The Bohannon experiment was selected for its applicability to natural science and because it is probably the largest and best documented experiment.

9 Critique and Variations of the Journal Peer Reviewing Process

9.1 Critique of the Peer Reviewing Process

During the 250 years since the invention of peer reviewing, the scientific community has continuously developed from a small group of scientists who knew each other individually to a globally acting industry, where colleagues even within one department may not know each other anymore. Peer reviewing has developed along, but at different speed, often haphazardly, and not based on empirical evidence. In recent years, concerns have been rising that peer review is completely ineffective (Jefferson et al. 2008)[1] or is not appropriate (anymore) for modern science.

Despite the frequent, sometimes rigorous and often fundamental critique on the peer reviewing system, the degree of acceptance is surprisingly high. Ware and Monkman (2008) showed that the majority of researchers (64% across fields) were satisfied with the reviewing process. In that study, 90% of researchers considered peer review as helpful for improving the quality of a paper. A more recent study by Mulligan et al. (2013) based on a larger dataset also showed a high degree of satisfaction with the peer reviewing system (69% over all fields in science with a range between 64% and 77% depending on the field); between 78% and 88% of the researchers supported the statement that the peer reviewing system greatly helps scientific communication.

---

[1]Jefferson et al. (2008) reviewed 28 studies that study the effect of peer review and found little empirical evidence to support the use of editorial peer review as a mechanism to ensure quality of biomedical research. They admit complex methodological problems in their study and cannot exclude that they simply failed to detect an effect. Unfortunately, only the negative statements were perpetuated and possibly contribute to the negative attitude towards peer review.

J.M. Starck, *Scientific Peer Review*, essentials,
https://doi.org/10.1007/978-3-658-19915-9_9

Despite the generally positive attitude towards peer review, critique is loud and explicit. Frequently presented points of critique on peer review are:

- It is too slow and too expensive.
- It is non-transparent.
- Peer review is insufficiently tested.
- It is not standardized.
- It is inhibitive to innovative ideas because reviewers tend to be more senior adhering to traditional thinking.
- It is unreliable because reviews frequently present divergent, often contrasting critique.
- It is unfair and biased against women, young researchers, and minorities.
- It prefers known scientists (Matthew effect[2]).
- It supports networks of established research groups.
- It does not acknowledge the work and intellectual input from reviewers.
- While the solicitation of reviewers' opinions by editors guarantees some comment and criticism for a manuscript, it will be unlikely that the editor will always have sent the manuscript to those particular researchers most knowledgeable and most capable of providing useful criticism.
- Peer review must fail because only reviewers close to the subject are knowledgeable enough to review, but these, being either competitors or friends, are disqualified by their conflict of interest (negative or positive, respectively).
- It cannot identify errors in data or detect research misconduct.
- The selection of peer reviewers may create problems because of a variety of reasons (bias, lack of experts in emerging and interdisciplinary areas, lack of experts due to the speed of research areas, etc.).

All these points are correct, but the central weakness of peer review that makes it fallible is that it is based on the honesty of individuals. Indeed, honesty represents a weakness because a minority of individuals acting incorrect can corrupt the

[2]"*For unto every one that hath shall be given, and he shall have abundance: but from him that hath not shall be taken even that which he hath.*" Matthew 25:29, King James Version (Merton 1968, 1988).

entire system, while the majority of researchers are honest and integer individuals.—In response to the critique, the community has experimented with different versions of the peer reviewing systems, alternative forms of quality control, and training, but there is no process that provides a measure of scientific quality without relying on honesty. In essence, there are only two approaches to the assessment of research: peer review and a metrics-based model (or a mixed model, combining these two approaches; Wilsdon et al. 2015).

9.2 Single Blind Reviewing

Single blind reviewing is probably the most frequently practiced editorial reviewing system. In the single blind reviewing process, the reviewers remain anonymous to the authors. The supposed advantage is that comments on a manuscript can be offered free and unrestrained by any hierarchical or personal dependency. The critique is that comments may consciously or unconsciously be biased in various directions (gender, hierarchical, national, minority), and that anonymity may open doors for unfair and derogatory comments. There is only little empirical support for the advantages as well as critique (see below). A study by Mulligan et al. (2013) reported that only 45% of researchers (n = 4037) considered single blind review as effective.

9.3 Double Blind Reviewing

In the double blind reviewing process, reviewer and authors remain anonymous to each other. In addition to the free and unrestrained comments from anonymous reviewers, an important supposed advantage is that reviewers cannot be biased because they do not know the authors. 76% of the researchers (across different fields in MINT) consider double blind review the most efficient (Mulligan et al. 2013). However, the empirical basis for this strong and positive support is surprisingly weak.

A frequent assumption is that the (single blind) peer review process is gender-biased, i.e., overall, men are treated more favorably than women. Double blind reviewing supposedly waives gender bias because names of authors are not visible to reviewers. The analysis of gender bias in science and in reviewing in particular has gained high attention because of its social relevance and because it can be analyzed relatively easily when surnames of authors are accessible. While there is no question about the existence of a gender gap in science, the analysis of its causes is more complicated because numerous factors contribute to it and because the prevalence of these factors has been changing over time. As a statistically significant development, the gender gap in science has slowly reduced during the past decade (Filardo et al. 2016; Helmer et al. 2017[3]) though there is still a considerable gap for functions as authors, reviewers, or editors. However, the question is if single blind reviewing contributes to creating/maintaining the gender bias and if double blind reviewing reduces it.—While Budden et al. (2008) showed that the double blind reviewing system effectively removes gender bias, a re-analysis of their data (Engqvist and Frommen 2008) could not confirm that double-blind review has any detectable positive effect. Later analyses, either on specific journal submissions (e.g., Primack et al. 2009; Buckley et al. 2014; Fox et al. 2016) or on large data-sets (e.g., Ceci and Williams 2011; Lee et al. 2013; Helmer et al. 2017), rendered no gender discrimination in reviewing (and final decisions by editors). Thus, recent analyses show a significant gender gap in science, but they provide no evidence that the reviewing process is contributing to it.

Fairness to unknown authors and improved quality of reviews have been listed as positive effects of double blind reviewing (review in: Snodgrass 2006). While it is already difficult to measure gender bias, bias in fairness and quality are even more problematic. It is, therefore, not surprising that some studies found proof of improved quality of double blind reviews (Okike et al. 2016), while others failed to find any effect (Alam et al. 2011).

The double blind reviewing system has been criticized because author identity cannot really be anonymized. For the specialist, it is simple to find out who the authors were. Also, some reviewers claim that knowledge of author identity can be positive in a sense that they value a manuscript that is obviously written by a young and unexperienced author differently, providing more support than if the manuscript was presented by an experienced senior researcher (although this could be considered positive bias).

[3]Based on the extrapolation of a large data set, Helmer et al. (2017) predict that exact parity could be achieved in 2027 for authoring, 2034 for reviewing, and 2042 for editing.

9.4 Open Peer Review

In open peer review,[4] authors and reviewers know each other. The system is supposedly transparent, fair, and confidential (to the general audience). Because reviewers and authors know each other, this prevents unfair or uninformative reviews. Some journals waive confidentiality and publish the reviews together with the paper so that the intellectual input and scientific contributions of the reviewers can be acknowledged by the community. While open peer review is transparent, it cannot prevent bias, Matthew effect, hierarchical submissive reporting, or incompetence.

9.5 Alternative Procedures—Post-Publication Peer Review

Transparency, sharing, disseminating, and receiving immediate feedback have become strong expectations not only in social media, but also in scientific publishing. The principle that ALL ideas should be discussed openly, debated, and archived, replaces the traditional selective process of eliminating the flawed and ill-conceived studies. With the advent of online-only journals, virtually unlimited space for publication, and the improved communication facilities through the internet, processes/workflows can be established that accommodate these expectations. Some modify or amend the traditional reviewing process, others aim at developing new tools of quality assessment to replace it.

An increasingly frequent practice is to publish articles on preprint servers before review, have the community comment on the article, and/or actively invite post-publication reviews.[5] The idea behind post-publication review is to establish priority, make scientific communication rapid, scientific discussion more public, community-based and transparent, and avoid editorial bias by selecting publications according to the journal's aims and scopes. Open (invited) peer review of articles is conducted after publication, focusing on scientific soundness and correct methodology rather than novelty or impact.

[4]In the current discourse the term "open peer review" refers to two different workflows: open final-version commenting (as discussed here) and open pre-review manuscripts are distinct phenomena (see below [9.6] for open pre-review publishing).

[5]As active post-publication reviewing sites see, e.g.: http://f1000research.com, https://www.scienceopen.com/, https://thewinnower.com, http://biorxiv.org/.

Preprint servers are not mutually exclusive with peer reviewing, they rather present an additional (new) form of presenting unpublished information (like conference talks etc.) for discussion and feedback by colleagues. Most preprint servers either invite reviews, allow reviewers to voluntarily select manuscripts, and invite editors to select reviewed manuscripts for publication in their journals. However, the post-publication reviewing systems in which the community comments and validates papers after publication, bears the potential to divert into a voting system, in which the majority of votes/comments consolidates paradigms instead of advancing science by recognizing that existing hypotheses have been rejected and new explanations may come in place. Open post-publication reviewing does not exclude bias and the number of positive comments may depend on the author's network. The competence of the commentators and the scientific soundness of their comments is not approved, and ultimately nobody knows who is reading the entire discussion associated with a paper.

Posting manuscripts on preprint servers potentially creates conflict with copyright when the paper is later published in a scientific journal. A comprehensive list of journals' attitudes towards preprints are available online and are regularly updated (http://www.sherpa.ac.uk/romeo/index.php).

9.6 The Experts and the Crowd

Science is based on a deductive system in which hypotheses are tested. Science proceeds by rejecting hypotheses and in a plethora of cases it was a single researcher who has done the right experiment documenting to the rest of the community that existing hypotheses were incorrect (e.g., Galileo, Kopernikus, Newton, Darwin, Einstein). Consequently, new explanations became established because the reproduction of the tests did not falsify the new hypotheses. Science is neither democratic nor a voting system in which the majority decides about validity. Because there is inertia in the community, the acceptance of new knowledge may be slow and it may require a long time before a correct explanation is accepted in sciences.—This categorical nature of scientific explanation has been quoted to support and reject the peer reviewing process. Indeed, there are good reasons for both views. (1) In support of the peer reviewing system, one may claim that innovative, new, and paradigm breaking ideas might only be recognized by the experts in the field and not by the crowd that tends to follow traditional views. Therefore, it has been argued, the peer reviewing system is a fair and good process to recognize and honor innovative and new ideas. (2) Opposing the peer reviewing system, some claim that exactly those experts may be conservative and persisting in traditional views, thus ignoring innovative, new, and paradigm breaking explanations in science. Also, an editor may have picked the wrong reviewer, who simply is incompetent. Only the wisdom of the crowd may ultimately recognize correct and better explanations.—However, it should be reiterated that science per se is not a democratic process, but it proceeds by falsifying hypotheses and replacing them by better explanations. Therefore, a majority voting system is inappropriate for research assessment.

9.7 The Role of Metrics in Research Assessment

The digitization of publishing has facilitated quantitative measures of publication impact. Citation-based bibliometric measures like impact factor and citation frequency are standard metrics. Recent years have seen the rise of "altmetrics" as a complement to those citation-based metrics. Altmetrics measure storage, links, bookmarks, and conversations associated with a paper, thus provide interesting additional information about how a paper is perceived by the community. Indeed, fifteen percent of researchers across fields in science believed that usage statistics instead of formal peer review would be effective (Mulligan et al. 2013).

A detailed survey of the role of metrics in research assessment (Wilsdon et al. 2015) concludes that the correlation between bibliometric indicators and peer review assessment varies between fields within the natural sciences. In some fields, citation-based indicators are good predictors of peer review outcomes, but in a number of fields there is no correlation. They recommend that it is currently not feasible to assess the quality of research output using exclusively quantitative indicators. In the Leiden Manifest that elaborates explicit recommendations for the correct use of bibliometric metrics in validating research, Hicks et al. (2015) state: "*Reading and judging a researcher's work is much more appropriate than relying on one number. [...] Research metrics can provide crucial information that would be difficult to gather or understand by means of individual expertise. But this quantitative information must not be allowed to morph from an instrument into the goal.*"—In the end, science is not a majority voting system.

10 Reviewer Recognition

Peer review developed as a collegial process, it was an integral part of being a scientist, and for the mutual benefit of all sides. It was a matter of course that scientists reviewed at least twice the number of manuscripts they had submitted to return to the community what they had received. With the industrialization of science, the exponentially growing number of manuscript submissions, and multiple manuscript submissions cascading down the impact ladder of journals, everybody is overloaded with work, the mutual benefit is less obvious, and many colleagues minimize the workload of reviewing because it is not recognized for their career. Thus, the balance between manuscript submissions and reviewer requests is in imbalance.

The last years therefore have seen a discussion about how to recognize the intellectual input and efforts to the progress of science contributed by the reviewers. If recognized as an important contribution to the community and a valuable point in the career of a researcher, reviewing might be more attractive. To maintain the balance between manuscript submissions and requested reviews, Fox and Petchey (2010) suggested a centralized "PubCred" bank that balances paper submissions by reviews. Submission of a manuscript would cost three PubCreds, while a completed peer review pays one PubCred. Authors would "pay" for their submissions with credits "earned" by doing reviews. Every individual would have an account held in the central "PubCred Bank."—A recent aberration of this mutual, collegial support has emerged as "zero-sum reviewing" in which researcher incur a 'reviewer debt' when publishing their own papers, and thereby feeling an obligation to pay back this debt by acting as a peer reviewer for other researchers' manuscripts. One standard algorithm for zero-sum reviewers is a simple $\Sigma k/n$ equation (cf Vines et al. 2010), where k is the total number of peer reviews received and n is the total number of co-authors on a paper (cumulatively summed over a body of work). At the heart of this idea is (still) the sentiment

J.M. Starck, *Scientific Peer Review*, essentials,
https://doi.org/10.1007/978-3-658-19915-9_10

that there is an obligation owed to the field. Zero-sum reviewing, however, rather minimizes individual responsibility and workload (Didham et al. 2017).

Other, new forms of reviewer recognition need to be established so that the work an individual researcher invests in the community is correctly and completely acknowledged. Indeed, reviews that improve the quality of a work and thus advance the field are not merely service to the community, but contributions to existing scholarship (Kennison 2016). Reviewer recognition, however, is to a certain degree in contrast to the frequently practiced anonymity of the reviewing process. Therefore, claims have been made that only open and transparent review processes allow for appropriate recognition of the reviewer's efforts.

10.1 Reviewer Recognition Platforms

Reviewer recognition has become an important topic and various attempts have been made how to develop a policy on peer review that takes into account the needs of authors, reviewers, and editors, and is flexible enough to deal with the editorial policies of the many journals, and still makes it possible for the research community to innovate. A successful, independent platform for reviewer recognition is *Publons*. The platform allows for various and flexible forms of reviewer recognition so that for the individual researcher peer review becomes a measurable indicator of expertise and contribution to their field.—*Publons* is a limited liability company registered in New Zealand and in the United Kingdom. It is free for reviewers, authors, and editors and finances itself by partnering with publishers. These partnerships allow reviews performed for cooperating journals to be automatically added to *Publons'* profiles.

Academic Karma is an Australia-based website that aims to make peer review free (as in freely accessible, as well as cost-free) and open for all academics. The website is linked with several preprint servers and addresses authors, reviewers, and editors with the aim of supporting open science. Reviewers can select manuscripts from preprint servers for open review. Reviewers maintain a single account for managing and recording all reviewing activity.

Reviewer Page is a publisher owned (Elsevier) public page listing all peer review activities carried out by a researcher. Elsevier-initiated, it aims to recognize the contribution of reviewers to scientific publications. The page records their review history for Elsevier journals and collects awards.

10.2 Who Owns the Review?

Copyright of an article is usually transferred to the publisher or the paper is published under a creative commons license. The copyright of the review is usually not transferred, thus it remains with the reviewer. This is straightforward and simple. However, with increasing industrialization of science, increasing productivity pressure on scientists, and the point-counting system for evaluations in job applications and promotions, researchers become more aware that the time they invest in the work of others, the efforts and intellectual achievement they contribute to the scientific community should be shared, recognized, discussed, and appropriately acknowledged. Therefore, some reviewers wish to publish their reviews in the internet, on personal websites, blogs, or more specialized sites.—However, the copyright of the reviewer on his/her review is in conflict with the confidentiality of the reviewing process as practiced by most journals. It is an "*industrial standard*" of the scientific publishing industry and an ethical request that reviewers should respect the confidentiality of the peer review process and not reveal details of the manuscript or its review beyond those released by the journal (COPE, ethical guidelines). Although only implicit with many journal guidelines, the general perception is that by accepting to review for a journal, the reviewer accepts to keep all communication about the manuscript and the review confidential. Journals should make this explicit so that no misinterpretations about confidentiality and of the reviewer's copyright can arise—generally, the community assumes priority of the commitment to confidentiality over the reviewer's copyright.

Institutions and Committees/ Recommendations in Conflict

11

11.1 Committee on Publication Ethics (COPE)

The Committee on Publication Ethics is a forum for editors and publishers of peer reviewed journals to discuss all aspects of publication ethics. It also advises editors on how to handle cases of research and publication misconduct. COPE offers useful resources also for authors and reviewers, including flowcharts and code of conduct (https://publicationethics.org/).

11.2 Council of Science Editors (CSE)

Science Editor is published quarterly in print and online by the Council of Science Editors (CSE), and also publishes weekly Early Online articles. It serves as a forum for the exchange of information and ideas among professionals concerned with editing and publishing in the sciences. Articles related to peer review studies, editorial processes, publication ethics, evaluating article impact, and other items of special interest to the journal's readers are encouraged (http://www.councilscienceeditors.org/).

11.3 EQUATOR Network

The EQUATOR Network (Enhancing the QUAlity and Transparency Of health Research) is an international initiative that seeks to improve the reliability and value of published health research literature by promoting transparent and accurate reporting and wider use of robust reporting guidelines (http://www.equator-network.org/).

J.M. Starck, *Scientific Peer Review,* essentials,
https://doi.org/10.1007/978-3-658-19915-9_11

11.4 European Association of Science Editors (EASE)

The European Association of Science Editors (EASE) is an international community of individuals and associations from diverse backgrounds, linguistic traditions, and professional experience in science communication and editing (http://www.ease.org.uk/).

11.5 International Committee of Medical Journal Editors

The ICMJE is a small working group of general medical journal editors whose participants meet annually and fund their own work on the Recommendations for the Conduct, Reporting, Editing and Publication of Scholarly Work in Medical Journals. The ICMJE invites comments on this document and suggestions for agenda items (http://www.icmje.org/).

11.6 Society for Scholarly Publishing (SSP)

The Society for Scholarly Publishing (SSP), founded in 1978, is a nonprofit organization formed to promote and advance communication among all sectors of the scholarly publication community through networking, information dissemination, and facilitation of new developments in the field (https://www.sspnet.org/).

11.7 International Society of Managing and Technical Editors (ISMTE)

ISMTE serves a unique niche within the academic, scientific, medical, technical, and professional publishing industry—editorial office staff. Through their newsletter, discussion forum, online resources, and meetings, they connect people in the profession. They provide networking and training infrastructure, establish best practices, and study and report on editorial office practices (http://www.ismte.org/?).

11.8 World Association of Medical Editors (WAME)

A global association of editors of peer-reviewed medical journals who seek to foster cooperation and communication among editors, improve editorial standards, promote professionalism in medical editing through education, self-criticism, and self-regulation, and encourage research on the principles and practice of medical editing (http://www.wame.org/).

11.9 World Medical Association: Declaration of Helsinki

The World Medical Association (WMA) has developed the Declaration of Helsinki as a statement of ethical principles for medical research involving human subjects, including research on identifiable human material and data (https://www.wma.net/policies-post/wma-declaration-of-helsinki-ethical-principles-for-medical-research-involving-human-subjects/).

11.10 Office of Research Integrity (ORI)

The Office of Research Integrity (ORI) is a US-American state agency that oversees and directs Public Health Service (PHS) research integrity activities on behalf of the Secretary of Health and Human Services with the exception of the regulatory research integrity activities of the Food and Drug Administration. It provides extensive resources to support research integrity, forensic tools to detect image manipulation, as well as overview on cases of research misconduct.

World Association of Medical Editors (WAME)

World Medical Association Declaration of Helsinki

Office of Research Integrity (ORI)

Take Home Message

- Peer review is the critical assessment of scientific reports by independent experts. It scrutinizes manuscripts, grant proposals or job applications for the correct application of the principles of science, correct scientific methodology, presentation according to the standards of scientific publishing, originality of research, and for legal and ethical correctness.
- Peer review has been established as a tool to recognize research that excels others, and to filter out science that was ill-conceived, poorly performed, or misinterpreted.
- Peer review depends on competence, trust, and goodwill of ALL participants. It makes an assumption of honesty and can assist in establishing scientific validity, but it cannot guarantee it—it is doomed to failure if the authors are dishonest or the reviewers incompetent.
- Scientific expertise is a prime responsibility of reviewers towards authors.
- Peer review is a collegial, thus social, process. It is founded in expert knowledge of the subject, methodology and literature relevant for specific decisions, but it is also inevitably subjective.
- Various versions of peer review exist today. All receive critique, but still the overwhelming majority of scientists is convinced that peer review helps to improve science and evaluate manuscripts by the quality of science.
- Its dependency on honesty and competence of the individuals involved in the reviewing process are strength and weakness of the system. To replace peer review, one would need to find a validating system that renders reliable results independent of the participants' honesty.

J.M. Starck, *Scientific Peer Review,* essentials,
https://doi.org/10.1007/978-3-658-19915-9

References

Alam M, Kim NA, Havey J, Rademaker A, Ratner D, Tregre B, West DP, Coleman WP (2011) Blinded vs unblinded peer review of manuscripts submitted to a dermatology journal. Br J Dermatol 165(3):563–567.

Benos, DJ, Bashari E, Chaves JM, Gaggar A, Kapoor N, LaFrance M., Mans R, Mayhew D, McGowan S, Polter A, Qadri Y, Sarfare S, Schultz K, Splittgerber R, Stephenson J, Tower C, Walton RG, Zotov A (2007) The ups and downs of peer review. Advances in physiology education, 31(2):145–152.

Bock W (2007) Explanations in evolutionary theory. J Zool Syst Evol Res 45:89–103.

Bock, W (2017) The dual causality and the autonomy of biology. Acta Biotheoretica 65:63–79.

Baker M (2016) Is there a reproducibility crisis? Nature 533:452–545.

Budden AE, Tregenza T, Aarssen LW, Koricheva J, Leimu R, Lortie CJ (2008) Double-blind review favours increased representation of female authors. Trends in Ecology and Evolution 23(1):4–6.

Csiszar A (2016) Troubled from the start. Nature 532:306–308.

Deakin L, Docking M, Graf C, Jones J, McKerahan T, Ottmar M, Stevens A, Wates E, Wyatt D, Joshua S (2014) Best practice guidelines on publishing ethics. © 2014 John Wiley & Sons, Ltd., CC BY-NC 4.0

Davies TG, Rahman IA, Lautenschlager S, Cunningham JA, Asher RJ, Barrett PM, Bates KT, Bengtson S, Benson RB, Boyer DM, Braga J, Bright JA, Claessens LP, Cox PG, Dong XP, Evans AR, Falkingham PL, Friedman M, Garwood RJ, Goswami A, Hutchinson JR, Jeffery NS, Johanson Z, Lebrun R, Martínez-Pérez C, Marugán-Lobón J, O'Higgins PM, Metscher B, Orliac M, Rowe TB, Rücklin M, Sánchez-Villagra MR, Shubin NH, Smith SY, Starck JM, Stringer C, Summers AP, Sutton MD, Walsh SA, Weisbecker V, Witmer LM, Wroe S, Yin Z, Rayfield EJ, Donoghue PC (2017). Open data and digital morphology. Proc R Soc B 284. doi: 10.1098/rspb.2017.0194

De Castro P, Heidari S, Babor TF (2016). Sex and gender equity in research (SAGER): reporting guidelines as a framework of innovation for an equitable approach to gender medicine. Ann Ist Super Sanità 52:154–157.

Engqvist L, Frommen JG (2008) Double-blind peer review and gender publication bias. Animal Behaviour, 76(3). 10.1016/j.anbehav.2008.05.023

J.M. Starck, *Scientific Peer Review,* essentials,
https://doi.org/10.1007/978-3-658-19915-9

European Commission (2008) EUR 23311—Mapping the maze: getting more women to the top in research Luxembourg: Office for Official Publications of the European Communities. ISBN 978-92-79-07618-3

Filardo G, da Graca B, Sass DM, Pollock BD, Smith EB, Martinez MAM (2016) Trends and comparison of female first authorship in high impact medical journals: observational study (1994–2014). BMJ 352:i847. doi.org/10.1136/bmj.i847

Fox J, Petchey OL (2010) Pubcreds: Fixing the peer review process by "privatizing" the reviewer commons. Bulletin of the Ecological Society of America 91:325–333.

Fox CW, Burns CS, Meyer JA (2016) Editor and reviewer gender influence the peer review process but not peer review outcomes at an ecology journal. Functional Ecology 30:140–153.

Garland Jr T, Adolph SC (1994) Why not to do two-species comparative studies: limitations on inferring adaptation. Physiological Zoology, 67:797–828.

Heidari S, Babor TF, De Castro P, Tort S, Curno M (2016) Sex and Gender Equity in Research: rationale for the SAGER guidelines and recommended use. Research Integrity and Peer Review 1:2. DOI 10.1186/s41073-016-0007-6.

Hicks D, Wouters P, Waltman L, de Rijcke S, Rafols I (2015) Bibliometrics: The Leiden Manifesto for research metrics. Nature 520:429–431.

Hochberg ME, Chase JM, Gotelli NJ, Hastings A, Naeem S (2009) The tragedy of the reviewer commons. Ecology Letters: 12: 2–4.

Ioannidis JPA (2005) Why most published research findings are false. PLoS Med 2(8): e124.

Merton RK (1968) The Matthew Effect in Science. Science. 159:56–62.

Merton RK (1988) The Matthew Effect in Science, II: Cumulative Advantage and the Symbolism of Intellectual Property. ISIS. 79:606–623.

Mulligan A, Hall L, Raphael, E (2013) Peer review in a changing world: an international study measuring the attitudes of researchers. Journal of the American Society for Information Science and Technology 64:132–161.

Paltridge B. (2017). The discourse of peer review. Reviewing submissions to academic journals. Palgrave McMillan/Springer Nature

Pan L, Kalinaki E. (2015) Mapping Gender in the German Research Area. Elsevier Analytical Services.

Popper K (1935) Logik der Forschung. Zur Erkenntnistheorie der Modernen Naturwissenschaft. Schriften zur Wissenschaftlichen Weltauffassung (P. Frank und M. Schlick, eds.) Bd. 9, Springer Verlag, Wien. Pp. 1–248.

Rennie D. (2012) Sense About Science 2012. CC-BY NC-ND 2.0. www.senseaboutscience.org

Rennie D (2003) Editorial peer review: its development and rationale. Peer review in health sciences, 2, 1–13.

Steinhauser G, Adlassnig W, Risch JA, Anderlini S, Arguriou P, Armendariz AZ, … Zwiren N. (2012). Peer review versus editorial review and their role in innovative science. Theoretical medicine and bioethics, 33(5), 359–376.

Wager E, Kleinert S (2011) Responsible research publication: international standards for authors. A position statement developed at the 2nd World Conference on Research Integrity, Singapore, July 22-24, 2010. Chapter 50 in: Mayer T and Steneck N (eds) Promoting Research Integrity in a Global Environment. Imperial College Press/World Scientific Publishing, Singapore (pp 309–16).

Walsh K (ed) 2016. Open Innovation, Open Science, Open to the World. A vision of Europe. Europen Commission, Brussels. doi:10.2777/061652

Ware M. and Monkman M. (2008). Peer review in scholarly journals: perspective on the scholarly community an international study. Publishing research consortium. Retrieved from: http://publishingresearchconsortium.com/index.php/prc-documents/prc-research-projects/36-peer-review-full-prc-report-final/file

West JD, Jacquet J, King MM, Correll, SJ, Bergstrom CT (2013). The role of gender in scholarly authorship. *PloS one*, *8*(7), e66212.

Wilsdon J, Allen L, Belfiore E, Campbell P, Curry S, Hill S, Jones R, Kain R, Kerridge S, Thelwall M, Tinkler J, Viney I, Wouters P, Hill J, Johnson B, (2015). The Metric Tide: Report of the Independent Review of the Role of Metrics in Research Assessment and Management. doi:10.13140/RG.2.1.4929.1363

Printed in the United States
By Bookmasters